YOUR KNOWLEDGE HAS VALUE

- We will publish your bachelor's and master's thesis, essays and papers

- Your own eBook and book - sold worldwide in all relevant shops

- Earn money with each sale

Upload your text at www.GRIN.com
and publish for free

Bibliographic information published by the German National Library:

The German National Library lists this publication in the National Bibliography; detailed bibliographic data are available on the Internet at http://dnb.dnb.de .

This book is copyright material and must not be copied, reproduced, transferred, distributed, leased, licensed or publicly performed or used in any way except as specifically permitted in writing by the publishers, as allowed under the terms and conditions under which it was purchased or as strictly permitted by applicable copyright law. Any unauthorized distribution or use of this text may be a direct infringement of the author s and publisher s rights and those responsible may be liable in law accordingly.

Imprint:

Copyright © 2018 GRIN Verlag
Print and binding: Books on Demand GmbH, Norderstedt Germany
ISBN: 9783668664678

This book at GRIN:

https://www.grin.com/document/416774

William Fidler

A possible future evolution of the universe

GRIN Verlag

GRIN - Your knowledge has value

Since its foundation in 1998, GRIN has specialized in publishing academic texts by students, college teachers and other academics as e-book and printed book. The website www.grin.com is an ideal platform for presenting term papers, final papers, scientific essays, dissertations and specialist books.

Visit us on the internet:

http://www.grin.com/

http://www.facebook.com/grincom

http://www.twitter.com/grin_com

A possible future evolution of the universe

W M Fidler

Abstract

The model of the future behaviour of the universe developed here is an extension of that developed in [1] and [2] from the indefinite past until the present time.

The first part of the trajectory of the deceleration parameter of cosmology into the future is obtained by extending the exponential variation with time of the trajectory from the end of the cosmic jerk until the present time, as developed in [1].

It is shown, in the future, when the magnitude of the deceleration parameter becomes, -1, that the universe attains a state where the volumetric strain rate becomes a minimum.

It is argued that, for this reason, the universe then remains in this state indefinitely.

The trajectory of the deceleration parameter of the universe from the present time into the indefinite future, as shown in the diagram in the body of the work, is then simply an exponential curve followed by a horizontal straight line.

Content

Introduction 4

The acceleration of space at the boundary of the universe 5

The future trajectory of the deceleration parameter 6

The special character of the state of the universe along the line q = -1 9

Discussion 12

References 13

Introduction

The predicted future state of any physical system is dependent upon the model of the system and the initial conditions. Since the veracity of the predictions from models at a given future time can only be determined by observation of the state of the system at that time, reckoned from some initial reference conditions, then none of the predictions can be considered to be definitive until that time is attained. Further, the quality of any model is crucially dependent upon the state of understanding of the underlying physics. The literature of cosmology, astrophysics and mathematics abounds with models predicting the state of the universe beyond the present time. Unless it can be shown that a certain model violates one or more of the physical laws, then the predictions from that model are equally valid in comparison with other models whose outcomes are diametrically-opposite.

It is in this spirit that, in Star Trek parlance, 'we boldly offer another model of the future state of the universe'.

The model is an extension of that of the evolution of the universe from its genesis until the present time as explained at length and developed in [1].

The acceleration of space at the boundary of the universe

It was shown in [2], p12, equation (9), that the linear acceleration of space, dv/dt, at the boundary of the universe is given by:

$$dv/dt = \sqrt[3]{N_{enc}} \cdot c/2v \left(dH/dt + H^2\right) \quad\text{------------ (1).}$$

Where N_{enc}, v and H are the number of enclosures (see [1] for an explanation), the frequency of the vacuum and Hubble's parameter, respectively.

It was noted in [2] that the bracketed term in the above expression may be shown to be equal to, $-qH^2$, where q is the deceleration parameter of the cosmic acceleration of the expansion of space in a FLRW universe. Hence, we may write equation (1) as:

$$dv/dt = -\sqrt[3]{N_{enc}} \cdot c/2 \left(qH^2/v\right) \quad\text{-------- (2).}$$

It is easy to see that the acceleration, denoted by α_0 at the present time and the acceleration denoted by α at some other time are then related by:

$$\alpha_0/\alpha = \left(qH^2/v\right)_0 \Big/ \left(qH^2/v\right) \quad\text{------------ (3).}$$

We have developed a series of equations, explained at length in [1] and [2], which have permitted the determination of q, H, and v. Without elaboration, we now repeat these expressions:

$$q = e^{k(t_0-t)} + B \quad\text{------------ (4).}$$

$$H/H_0 = 1 \Big/ \left\{1 - H_0/k \left[q_0\omega + e^\omega - 1\right]\right\} \quad\text{------------ (5),}$$

Where $\omega = k(t_0 - t)$.

$$v/v_0 = \exp\left[\int_t^{t_o} H\, dt\right] \quad \text{------------------------------ (6)}$$

The future trajectory of the deceleration parameter

It was shown in [1] that the trajectory of the deceleration parameter from the time of the end of the cosmic jerk until the present time, under the assumptions shown there, was given explicitly by:

$$q = \exp\left(0.8160679 * 10^{-18} * (t_0 - t)\right) - 1.4 \quad \text{------------------------- (7).}$$

We now extend this trajectory into the future, not, it should be emphasised, indefinitely.

Now, Hubble's parameter, **H** may be expressed as: $H = \frac{1}{a}\,da/dt$ ---------------- (8), where **a** is the scale factor.

If this expression is differentiated wrt t, we get: $dH/dt = \frac{1}{a}\,d^2a/dt^2 - \frac{1}{a^2}\left(da/dt\right)^2$,

Which may be written: $dH/dt = \frac{1}{a}\,d^2a/dt^2 - H^2$ -------------------- (9).

The deceleration parameter, q, is defined as: $q = -a \Big/ \left(da/dt\right)^2 \cdot d^2a/dt^2$ ----------- (10).

Combining (9) and (10) we obtain: $\frac{1}{H^2}\,dH/dt + (1 + q) = 0$ -------------------- (11).

Examination of (11) shows that, if **q** exceeds -1 then, dH/dt becomes positive, and hence, thereafter, the Hubble parameter increases with the passage of time.

We cannot advance any plausible argument in support of an increasing Hubble parameter with time and hence we take $q = -1$ to be the absolute negative limit of the deceleration parameter. Indeed, if the point in time at which **q** becomes equal to -1 is denoted by the symbol t_{0+}, the subscript '0+' denoting a new time origin in the future, then if **q** is taken have the magnitude, **-1** from this origin into the indefinite future, then, for all time beyond this origin, dH/dt will always be zero, **H**, constant, and the minimum value of Hubble's parameter from now into the indefinite future.

Denoting this constant, and minimum, Hubble parameter as H_{0+}, then we may write equation (9) as:

$$d^2a/dt^2 - H_{0+}^2 a = 0 \text{ -------------- (12)}.$$

We now have a simple, second-order, constant coefficient, linear differential equation which describes the variation of the scale factor with time along a line, $q = -1$, at all times in the range, $t_{0+} \leq t_+ \leq \infty$. Where t_+ denotes any time in the future of t_{0+}.

For consistency we re-write equation (11) as follows:

$$d^2a_+/dt_+^2 - H_{0+}^2 a_+ = 0 \text{ -------------- (13)}.$$

Two solutions are: $a_+ = \xi_1 \exp(H_{0+} t_+)$, and, $a_+ = \xi_2 \exp(-H_{0+} t_+)$.

As shown in [2], equations of this form may be rewritten:

$$a_+ = \xi_1 \exp(H_{0+}(t_+ - t_{0+})) \text{ and } a_+ = \xi_2 \exp(-H_{0+}(t_+ - t_{0+})).$$

At $t_+ = t_{0+}$, $a_+ = a_{0+}$, hence $\xi_1 = \xi_2 = a_{0+}$.

We discard the latter of the two solutions for it indicates that the scale factor decreases with time. Further, it may be inferred from equation (5) of [2] that, for this model, a decreasing scale factor would be associated with an increase in the frequency of the vacuum, for which we have no explanation. Moreover, the latter solution is incompatible with equation (8).

Hence, the variation of the scale factor with time from the origin '0+' is taken to be:

$$a_+ = \xi_1 \exp(H_{0+}(t_+ - t_{0+})) \text{ -------------- (14)}.$$

Now, at $t_+ = t_{0+}$, $a_+ = a_{0+}$, $\therefore \xi_1 = a_{0+}$.

Equation (14) may then be written:

$$a_+ = a_{0+} \exp(H_{0+}(t_+ - t_{0+})) \quad \text{---------------- (15).}$$

Now, we can determine the time, t_{0+} at which the new origin is located with respect to the present time by setting **q = -1** in equation (7).

Hence, $t_{0+} = (t_0 - t) = \dfrac{10^{\wedge}18}{0.8160679} \ln 0.4 = -1.1228 * 10^{18}$ s.

The negative sign indicates that this time is in the future of t_0 and is equivalent to **35.6 billion years** from now.

It is shown in [2] that the Hubble parameter, for the distribution given by equation (7), is expressed by:

$$H = \dfrac{H_0}{\{1 - H_0[(1+B)(t_0 - t) + (e^{k(t_0 - t)} - 1)/k]\}} \quad \text{-------- (16).}$$

Here, $H_0 = 2.177 * 10^{-18}$ s^{-1}, $k = 0.8160679 * 10^{-18}$ s^{-1}, $B = -1.4$.

In conjunction with this data and the accompanying time interval, then the magnitude of the Hubble parameter at '0+' is calculated to be $H_{0+} = 1.3415 * 10^{-18}$ s^{-1}.

The following diagram, which is not to scale, replaces the curved path of the trajectory of **q** from the present time to the position '0+', with a straight line, and from where is drawn the remainder of the trajectory which is a true straight and horizontal line; the combination of lines represents the overall trajectory of the deceleration parameter from now,'0', into the indefinite future.

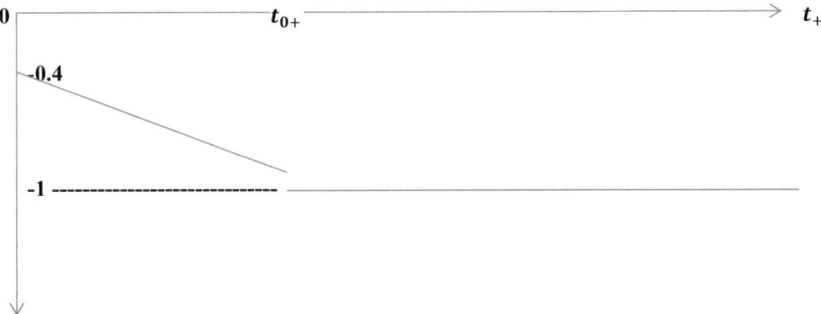

It was shown in [2] that the equivalent blackbody radiation temperature, θ, was related to the vacuum frequency, ν, by the expression: $\theta/\nu = $ **a constant**.

Hence, $(\theta/\nu)_0 = (\theta/\nu)_{0+}$, from which it follows that $\theta_{0+} = \theta_0 \, {}^{\nu_{0+}}/\nu_0$.

The frequency ratio as determined by the method shown in [2] is **0.1758** and, in conjunction with the currently-accepted value of the blackbody radiation temperature at the present time, viz. **2.725 K,** then the equivalent blackbody radiation temperature, θ_{0+} at time t_{0+} is calculated to be **0.479K**.

The special character of the state of the universe along the line q = -1

It was shown in [2] that the Hubble parameter could be assigned a meaning other than a proportionality factor relating the speed of expansion of space at a point to the distance between that point and an observer.

Independent of any particular dynamic of the universe it was shown there that the rate at which space was being created for the whole universe ('**u**') is given by:

$$dV_u/dt = 3 V_u H \quad\text{------------------------------ (17).}$$

This may be rearranged as: $\left[\frac{dV_u/V_u}{dt}\right] = 3H$ -------------------- (18).

We may interpret the LHS of this expression as the volumetric strain rate of the universe. Hence, along a line $q = -1$ where, $H = H_{0+}$, and in view of the foregoing discussion, it may thus be inferred that the volumetric strain rate of the universe is a minimum, and hence, the universe may reasonably be expected to remain in this state indefinitely.

Therefore, $\left\{\left[\frac{dV_u/V_u}{dt}\right]\right\}_{min} = 3H_{0+}$ -------------------- (19).

The magnitude of the minimum strain rate of the universe is hence, **4.0245 * 10^-18 s^{-1}**.

Along $q = -1$ the acceleration ratio as expressed by equation (3) may be written:

$$\alpha_{0+}/\alpha_+ = \frac{\left(qH^2/v\right)_{0+}}{\left(qH^2/v\right)_+}$$ -------------------- (20).

But, $(qH^2)_{0+} = (qH^2)_+$, it then follows that, $\alpha_{0+}/\alpha_+ = v_+/v_{0+}$.

Now, as shown in [2], $v/v_0 = l_0/l = a_0/a$.

Along $q = -1$, $v_+/v_{0+} = \alpha_{0+}/\alpha_+$, which, from (15) equals $exp(H_{0+}(t_{0+} - t_+))$

Hence, $\alpha_{0+}/\alpha_+ = exp(H_{0+}(t_{0+} - t_+))$ -------------------- (21).

Since $H_{0+}(t_{0+} - t_+)$ is negative, then α_+ is greater than α_{0+}, increasingly so as t_+ increases.

We may now write down the complete set of equations along the line **q = -1** from which the state of the universe may be determined, from that time, (t_{0+}), at which **q** first becomes equal to **-1**, into the indefinite future, i.e.

$$v_+/v_{0+} = l_{0+}/l_+ = a_{0+}/a_+ = \alpha_{0+}/\alpha_+ = exp(H_{0+}(t_{0+} - t_+)) \text{ ---------------- (22)}.$$

Where, as calculated previously, $H_{0+} = 1.3415 * 10^{-18} \ s^{-1}$.

Now, $v_0/v_{0+} = l_{0+}/l_0 = a_{0+}/a_0 = \alpha_{0+}/\alpha_0$. The inverse of the first of these ratios, viz.

0.1758, has been determined earlier; as is customary, a_0 is set equal to unity, and hence it is calculated that magnitude of the reference scale factor, a_{0+} at the beginning of the phase, $q = -1$, is, $a_{0+} = 5.688$.

Using Penrose's assumption [3], of a universe containing 10^{80} baryons and using the baryon matter density as determined from the WMAP survey [4] it was shown in [1] that the number of enclosures corresponding to this number of baryons is, $2.628 * 10^{92}$, and is constant.

As shown in [1], the values of, q_0, H_0, and v_0 are, -0.4, $2.177* 10^{-18} \ s^{-1}$ and **2.681 T Hz** respectively.

Hence, from equation (2) it is calculated that at the boundary of the universe, space is accelerating, currently, at a rate of, $0.679 * 10^{-9} \ m^2/s$. Using the above formulae, the reference acceleration of space, α_{0+}, at the boundary of the universe is determined to be:

$\alpha_{0+} = 3.862 * 10^{-9} \ m^2/s$.

In summary, the reference conditions at the beginning of the phase, **q = -1** are:

$v_{0+} = 0.471 \ T \ Hz$, $a_{0+} = 5.688$, and, $\alpha_{0+} = 3.862 * 10^{-9} \ m^2/s$, respectively.

Discussion

The model of the universe presented here is based upon that developed in earlier work and explained at length in [1]. In that model plausible explanations are developed for the genesis of the dark and baryonic matter, the formation of the primordial light elements, the surface of last scattering and the cosmic jerk. All of the foregoing is attributed to a single mechanism. The extension into the future of the trajectory of the deceleration parameter consists simply of an exponential curve followed by a horizontal straight line. The criterion for the shape of the second portion of the trajectory is obtained from equation [11] and the choice is justified subsequently, by showing that at all points along this line, the magnitude of the volumetric strain rate of the universe is a minimum.

The model of the universe developed here and in [1] and [2] has an actual boundary at any given time for it has developed from an expanding cubic bubble of space of indefinitely small size in the indefinite past. As noted in [2] the inception of the universe developed further in [1] requires no knowledge of its antecedents, other than the postulate that, prior to inception, the universe was, at all times, of a finite size and, on a net basis, absolutely empty. For the purpose of illustration the size of the universe at the 'Planck Point' is now calculated. Using Planck's 'h', then when the oscillators of this model are vibrating at the Planck Angular Frequency, $v_P = 7.399825 * 10^{42}\ Hz$, the volume of a single enclosure is the cube of the Planck Wavelength, i.e. $(c/v_P)^3 = 66.6345 * 10^{-105}\ m^3$. Now, for a universe currently containing the Eddington number of baryons, i.e. 10^{80} we have shown previously that the number of enclosures, $N_{enc} = 2.628 * 10^{92}$. The volume of this universe, V_{UP}, when the vacuum is oscillating at the Planck Frequency, is given by the product of the number of enclosures (which is taken to be constant) and the volume calculated above. Hence, the volume of what here may be termed the Planck Universe, $V_{UP} = 17.5115 * 10^{-12}\ m^3$.

This may be represented by a cube of side, $2.6 * 10^{-4}\ m$.

Prosaically, we may say that, for this model, the volume of the universe, shown elsewhere, [2], to be represented at the present time by a cube of side **78 billion light years**, at an unspecified time before the accepted time of inception, was, roughly, no larger than a grain of sugar (facetiously, the Planck Grain).

References

[1] Cosmological Heresies – a new model of the evolution of the universe.

W M Fidler, November 2017

Submitted to GRIN for publication.

[2] The structure and expansion of a universe obeying Hubble's law and whose vacuum is modelled as a set of identical, bi-modal, frequency-quantised, simple quantum harmonic oscillators.

W M Fidler, February 2016.

Unpublished.

[3] The Road to Reality.

R Penrose

Jonathan Cape 2004.

[4] Hyperphysics-phy-astr.gsu.edu/hbase/Astro/wmap.html

YOUR KNOWLEDGE HAS VALUE

- We will publish your bachelor's and master's thesis, essays and papers

- Your own eBook and book - sold worldwide in all relevant shops

- Earn money with each sale

Upload your text at www.GRIN.com and publish for free